Emerging Technologies in Clean Energy

Topical Conference at the 2022 AIChE Spring Meeting and 18th Global Congress on Process Safety

San Antonio, Texas, USA
10 – 14 April 2022

ISBN: 978-1-7138-6817-0

Printed from e-media with permission by:

Curran Associates, Inc.
57 Morehouse Lane
Red Hook, NY 12571

Some format issues inherent in the e-media version may also appear in this print version.

Printed with permission by Curran Associates, Inc. (2023)

For permission requests, please contact AIChE
at the address below.

AIChE
120 Wall Street, FL 23
New York, NY 10005-4020

Phone: (800) 242-4363
Fax: (203) 775-5177

www.aiche.org

Additional copies of this publication are available from:

Curran Associates, Inc.
57 Morehouse Lane
Red Hook, NY 12571 USA
Phone: 845-758-0400
Fax: 845-758-2633
Email: curran@proceedings.com
Web: www.proceedings.com

TABLE OF CONTENTS

638748 Application of Bowtie Analysis to Sub-Surface CO_2 Disposal

Abstract:

Bowtie analysis is an established barrier-based risk assessment technique[1] that involves building a pictorial representation of plausible accident scenarios – a bowtie diagram – usually in a multi-disciplinary workshop environment. It provides a means of analysing, qualitatively and in depth, the possible causes of an unwanted event and the potential consequences should such an event occur and/or escalate. It also enables an evaluation of the individual prevention and mitigation measures – the barriers.

In the context of CO_2 disposal, bowtie analysis has become a standard technique for identifying and analysing the risks of leakage from geological CO_2 storage sites.

Bowtie diagrams can be understood at all levels, including by non-risk specialists. It is, therefore, a convenient tool for communicating various aspects of a particular hazard and the means by which it is managed in a clear, logical, and substantiated manner. This is especially beneficial in an emerging field such as sub-surface CO_2 storage, where the level of risk perceived by the public can be significant, and the nature of the risks and how they are managed can be highly technical. Bowties are very effective in enabling diverse stakeholders to visualise how risks are being managed.

The process of developing a bowtie diagram also brings significant benefits, as the bowtie workshop provides an opportunity for all parties involved in the CO_2 storage project to raise and discuss potential problems, allowing each discipline to appreciate how their decisions and plans can impact on other areas of the project. By aiding a clearer, shared understanding of all aspects of how a particular risk scenario can occur and how it is managed, the bowtie exercise contributes directly to more informed, and better, decisions being made.

Application of bowtie analysis within the oil and gas industry globally is mature; it has been in widespread use since the mid-1990s and tends to be used to illustrate how Major Accident Hazards (MAHs) are managed. In this context, MAHs include hydrocarbons being extracted, processed, stored and distributed, and the unwanted events usually relate to loss of containment, fires, explosions, etc. with potentially catastrophic consequences. Prevention and mitigation measures are typically engineered systems and operating limits.

By drawing on real-life practical applications[2,3,4,5] and the findings from multi-year research projects[6,7], this paper presents the challenges, and some solutions, to applying the bowtie analysis technique to sub-surface storage of CO_2, where the unwanted event is similarly loss of containment but, unlike the analysis of process plant hazards, the prevention and mitigation measures include natural geological formations which act as impermeable containment barriers.

While bowties are always subjective, the absence of conventional process industry-type reliability data for geological barriers means the analysis must adopt different ways of evaluating the barriers and accounting for uncertainty. The range of causes which can lead to CO_2 loss of containment from sub-surface storage, and the way in which these causes (or pathways) are interconnected, means similarly interconnected bowtie diagrams may be necessary. Finally, proving that the risk of unwanted CO_2 release from storage is adequately managed can be challenging, in the absence of numerical risk values and acceptability criteria.

This paper presents an introduction to bowtie analysis and references relevant case studies to illustrate how the analysis can be adapted for application to sub-surface storage of CO_2. It concludes that bowties are well suited for detailed, structured qualitative understanding of the risks of loss of CO_2 containment, with potential for some (limited) quantification.

References

1. Center for Chemical Process Safety of the American Institute of Chemical Engineers, in association with the Energy Institute. *Bowties in Risk Management, a Concept Book for Process Safety*
   Wiley, ISBN 978-1-119-49039-5, 2018.
2. Northern Lights - Receiving and permanent storage of CO_2 Plan for development, installation and operation, Part II - Impact Assessment,
   EL001 October 2019. https://northernlightsccs.com/en/facts-and-reports
3. ENI UK Ltd. Liverpool Bay Carbon Capture and Storage Project
   https://www.eni.com/en-IT/operations/united-kingdom-liverpool-bay.html
4. Northern Endurance Partnership, Endurance Carbon Capture and Storage Project
   https://www.netzeroteesside.co.uk/
5. https://www.portofrotterdam.com/en/news-and-press-releases/ccs-project-porthos-a-step-closer
6. Risktec Solutions BV. Bowtie Analysis Report: DETECT: Integrated CO_2 Leakage Risk Assessment,
   Report No. SGSI-12-R-07, Issue 1, 2020.
7. SECURe project (Subsurface Evaluation of CCS and Unconventional Risks).
   https://www.securegeoenergy.eu/work-packages/wp2-risk-assessment

639245 Renewable Energy to Synthetic Natural Gas

Abstract:

The utilization of renewable energy and carbon capture have become popular and essential topics in the energy industry during the past years due to increasing concerns about global warming. This paper studies the Sabatier reaction for methanation of CO2 from a fossil fuel power plant as a process step in the production of Synthetic Natural Gas (SNG) utilizing renewable energy. The process is modeled using CHEMCAD process simulation software.

When renewable energy is used for electrolysis of water to produce hydrogen and combined with the Sabatier reaction, there is a pathway from renewable energy to SNG. The resulting SNG may be recycled back to the process resulting in near zero emissions or used elsewhere as a product.

639315 Analyzing CO_2 Hydrogenation Activity of Ga-Cu Alloys By Experimental and Computational Methods

Abstract:

Climate change is an environmental curse caused by excessive CO_2 releases in the atmosphere caused by the anthropogenic combustion of fossil fuels. Researchers currently focus on making value-added chemical products from CO_2 such as methanol, DME, formic acid, etc., on reducing greenhouse gases (GHG) and ameliorate global warming. The catalytic hydrogenation of CO_2 represents one viable option for utilizing CO_2. In recognition of the growing need for methanol (produced mainly from syngas) and the global effort to reduce greenhouse gas emissions, researchers have a terrific opportunity to convert CO_2 captured from a wide range of sources using heterogeneous catalysts. Catalysts commonly used in CO_2 hydrogenation include Cu, Pd, In, Ni, and bimetallic catalysts (Cu-Zn, Pd-In, and Ga-Cu). Several metals supported on oxides have shown promise as catalysts for CH_3OH synthesis from CO_2. Despite the wide use of oxide-supported metals as catalysts for CO_2 hydrogenation reactions, very little comprehensive knowledge of the mechanisms and key steps that contribute to activity and selectivity. Creating a better catalyst by design depends on the understanding of atoms at the atomic level. Towards that goal, we carried out the first principle density functional theory calculations for the major reactions that can form methanol and DME from CO_2.

Experiments were conducted in 16-channel Flowrence XD from Avantium at temperature of 225-300°C, 50 bar pressure and CO_2 to H_2 ratio of 1:4. Vienna Ab Initio Simulation Package (VASP) was used to calculate the reaction kinetics using the first principle density functional theory (DFT). The climbing image nudged elastic band method (CI-NEB) was used to identify the transition state structures of the elementary reactions. The following equation was used to calculate the binding energy of all the species present in CO_2 hydrogenation: $E_{bind} = E_{adsorbate+surface} - E_{adsorbate} - E_{surface}$

where $E_{adsorbate+surface}$ is the total energy of the adsorbate on metal surface, $E_{adsorbate,}$ and $E_{surface}$ are the total energy of adsorbate in the gas phase and the bare metal surface

Research has been devoted to the development of catalysts that can promote the methanol synthesis process from CO_2 hydrogenation. It has recently been found experimentally that a Ga-Cu intermetallic catalyst exhibits high activity, selectivity, and stabilities for this reaction. We present here a combined density functional theory and kinetic modeling study in order to provide insight into the reactions related to methanol and dimethyl ether (DME) formation over Ga-Cu (111). The activation barriers and adsorption energies of the major primaries in CO_2 hydrogenation reaction mechanism have been calculated via DFT calculations on Ga-Cu (111) surfaces. Many reaction intermediates are involved in the CH_3OH production from formate and carboxyl reaction pathways, and it can occur in various ways. An understanding of the active intermediates in CO_2 hydrogenation and their role in CH_3OH formation can only be achieved with experimental observations and theoretical calculations combined. We demonstrate that methanol can be synthesized using HCOO* as well as COOH* reaction pathway rather than just HCOO* over various catalyst surfaces. We found that the HCOO* formation is possible reaction pathway for the formation of methanol, because energy barrier is lower than the carboxyl pathway. Further observed that the formation of methanol via hydrogenation of CH_3O is kinetically favorable compared to the hydrogenation of CH_2OH. This analysis provides insight into the role surface structure play in defining the nature of Ga-Cu catalysts used in DME and methanol syntheses. Experimental results showed that the selectivity of methanol was higher in equal percentages of Ga and Cu, but the selectivity of CO was higher in Ga-rich surfaces. There is an interesting fact that the DME can only be observed with addition of Ga to the Cu. According to the DFT calculations, the Cu rich surface had the lowest activation energy for methanol formation, and the Ga rich surface had the lowest activation energy for DME formation. Experimental observations and DFT results are in good agreement. By applying microkinetic modelling techniques to the calculated activation barriers presented in the current study, future studies will be able to better understand the mechanism of methanol and DME synthesis and the relative importance of the identified intermediates.

639325 Techno-Economic Analysis of the Effects of Biomass Co-Firing and Clc Oxygen Carriers on the Performance of Coal Based IGCC Power Plant

Abstract:

The global demand for energy is constantly growing due to the population and economy growth, especially in emerging market economies. Fossil fuels still produce most of global energy with coal being the most abundant fossil fuel. The use of coal has shown an increase over time, which resulted in an increase in carbon dioxide emissions. Even with the rise in the global awareness about the negative effects caused by greenhouse gases, the abundance and the low cost of coal still ensures the existence of coal-fired power plants.

This research the potential of retrofitting integrated gasification combined cycle (IGCC) power plants with biomass co-firing and chemical looping combustion (CLC) carbon capture for near zero carbon emissions. The investigated biomasses are lignin; palm coir, palm empty bunch and palm leaflets (agrowastes in UAE). The effect of different biomass co-firing and CLC oxygen carrier on the techno-economic performance of coal-based IGCC power plants was modelled and simulated using Aspen Plus software.

The results showed that the use of iron oxide as an oxygen carrier better performance than the traditional oxygen carrier of nickel oxide in terms of total plant efficiency, total plant energy output and carbon specific emissions. However, to achieve near zero carbon emissions, the flow rate of the oxygen carriers was 15 times the flow rate of the feed.

Increasing the percentage of lignin and palm coir co-firing resulted in better performance regardless of the used oxygen carrier. However, the performance has diminished when the percentage of palm empty bunch and palm leaflets co-firing increased. One of the major drawbacks of biomass co-firing is the particulate matter that remains. The presence of particulate matter results in downstream equipment fouling and hence, lowering the overall power plant efficiency. This issue has been addressed by adding cyclones to the syngas cleaning section.

The use of CLC for carbon capture has increased the cost of power production and the increase depended on the type of oxygen carrier. The cost of power production has increased by 32% when iron oxide was used, whilst the cost has increased by 45% when nickel oxide. However, there was no clear trend to indicate if increasing biomass co-firing percentage will either increase or decrease the cost of power production. Unfortunately, the lowest cost was greater than actual market trading prices. Therefore, currently no clear economic drive for developing coal biomass co-firing power plants with CLC carbon capture technology.

Despite everything, a satisfactory coal and biomass co-firing power plant with suitable decarbonisation scenario might be economically feasible, depending on a trade-off between the decarbonisation degree and CO_2 specific emissions. All that will be driven by the introduction of new national/international legislation regarding allowable carbon emissions from coal-fired power plants.

639340 Copper-Exchanged Zeolitic Imidazolate Framework-8 Enable Selective CO_2 Hydrogenation to Methanol

Abstract:

Introduction

In view of mitigating CO_2 emissions and cut the dependency of our reliance on fossil feedstocks, number of approaches were proposed to capture and convert CO_2 to useful fuels or chemicals. While methanol (MeOH) synthesis via CO_2 hydrogenation might become practically relevant in the future, many researchers working on to develop an active, selective, and stable catalyst with low-cost. Currently, a copper-based catalyst is industrially used to produce methanol from a mixture of $CO/CO_2/H_2$ at 200–300 °C and elevated pressures (50 – 100 bar) since 1960s. Despite high methanol productivity over the benchmark $Cu/ZnO/Al_2O_3$ catalyst, this catalytic system suffers with faster deactivation and selectivity issues (reverse water gas-shift reaction, and methanation). To address the issues with selectivity and stability of the catalyst, a number of multimetallic composite systems have been proposed, that is, $CuO/ZnO/ZrO_2$, supported (e.g. ZnO, SiO_2, ZrO_2) Pd, and In-Pd, $Au/CeO_x/TiO_2$, Ni-Sn/In-ZrO_2, Ni/In_2O_3, Ni-Ga/SiO_2, In_2O_3/Co_3O_4, and etc. However, despite they achieve high selectivity towards methanol with better stability, these mix-component systems are able to convert CO_2 to methanol at low conversions per pass and/or the high cost of some of the materials used in making the catalysts did not see further scope for scale-up and commercialization. In this reference, we explore the synergistic interaction of copper stabilized Zn-MOF (ZIF-8) for CO_2-based methanol synthesis through a comprehensive experimental and theoretical program. *In situ* IR spectroscopy and surface adsorption techniques along with thermal, volumetric, and microscopy analyses uncovered the nanometric exchange and construction of the selective copper species in ZIF-8 framework. Density functional theory used to understand structural rearrangements of copper exchanged ZIF-8 and the reactivity of complementary promoted surfaces.

Materials and methods

The ion-exchange and impregnation procedure for ZIF-8 sample was carried out with excess metal nitrate (in methanol or water) solution at 50-80 °C. After evaporation-exchange procedure, the sample was dried at 100 °C for 24 h prior to any pre-treatment, characterization, and catalytic testing of the material. Textural properties were studied by means of N_2 adsorption-desorption (Micromeritics ASAP 2040) at -196 °C. Crystallographic analyses were performed via powder X-ray diffraction in a Bruker D8 advanced diffractometer equipped with a Bragg-Brentano geometry fitted with a copper tube operating at 40 kV and 40 mA. Diffractograms were acquired over a 2θ range of 10-80°. Metal-support interactions were studied by means of temperature-programmed H_2 reduction (Altamira AMI-200ip). Chemical composition of the powdered samples was analyzed by ICP-OES (Varian, Inc./Agilent Model 7200-ES) and high resolution X-ray photoelectron spectroscopy in a Kratos Axis Ultra DLD spectrometer equipped with a monochromatic Al Kα x-ray source (hv = 1486.6 eV). High-pressure *In situ* Diffuse Reflectance FT-IR-Spectroscopy (DRIFTS) measurements were performed on a Nicolet 6700 FTIR spectrometer with a liquid-nitrogen cooled MCT detector using a high-temperature reaction cell (Harrick) equipped with a temperature programmer and connected to a gas-dosing device. Scanning electron microscopy imaging was performed with secondary electrons in an

FEI TENEO VS microscope using 2 kV acceleration voltage and 5 mm working distance. Catalytic tests were performed using a Flowrence® Avantium parallel reactor consisting of 16 tubular fixed-bed quartz reactors. In each reactor, ~ 0.05 g of sieved catalyst particles (150 - 300 mm) were loaded onto a 9.5 cm long coarse SiC bed that ensures the catalyst bed lies on the isothermal zone of the reactor. The reactors were pressurized with a mixed feed containing 20 vol % of CO_2 and 80 vol % of H_2 to 50 bar using a membrane-based pressure controller in the temperature range of 225 – 300 ºC and the effluent stream was analyzed using a gas chromatograph equipped with TCD and FID detectors.

Results and discussions

The prepared catalysts were extensively characterized by various adsorption and spectroscopic techniques to understand the surface, bulk, and morphological properties. The elemental composition of all the samples were investigated by ICP-OES, STEM-EDX, and XPS analyses and percentages of each element determined from ICP-OES analysis are in line with the nominal loadings. A higher BET surface area observed for ZIF-8 sample (~1160 m^2/g) decreased significantly after Cu exchange or impregnation. However, Cu-exchanged ZIF-8 sample showed much higher surface area (~870 m^2/g) compared to the Cu impregnated ZIF-8 (~25 m^2/g) and benchmark Cu-ZnO-Al_2O_3 (~78 m^2/g) catalysts. The thermogravimetric analysis of Cu ion-exchanged catalyst showed relatively better thermal stability in comparison to Cu impregnated sample. The structural stability of ZIF-8 MOF after Cu-exchange is also confirmed with *in situ* diffuse reflectance infrared spectroscopy at reaction temperature (250 °C) under high pressure (25 bar). In reference to the testing of these materials in CO_2 reduction methanol, a higher space-time-yield was observed for Cu exchanged MOF in comparison to the impregnated and benchmark catalysts. The higher productivity observed for Cu-exchanged catalyst was attributed to the presence of higher CO_2 sorption capacity (CO_2-TPD), uniform distribution of Cu nanoparticles (SEM and HR-TEM), improved thermal stability (TGA and XRD) caused after Cu exchange. The higher thermal stability of the Cu-exchanged ZIF-8 catalyst was attributed to the increase in the ZnO crystallite size determined from XRD analysis of the catalyst. The high-pressure *in situ* diffuse reflectance infrared spectroscopic studies of the catalysts revealed the possible reaction intermediates in order to understand the reaction mechanism. Density functional theory is also used to understand the gas (CO_2+H_2) adsorption properties and reaction mechanism of CO_2 hydrogenation to methanol. Finally, in addition to the high activity and selectivity of the prepared catalytic system, the catalyst also showed good stability for more than 36 h of continuous operation in CO_2 to methanol reaction conditions. The used MOF catalyst was also characterized by XRD, XPS and HR-TEM analyses that showed insignificant differences in comparison to the fresh catalyst indicating its robust nature under the reaction conditions used in this study.

Conclusions

A detailed kinetic, mechanistic and theoretical investigations were carried out for a newly synthesized MOF material in the CO_2 hydrogenation to methanol. The investigations revealed superior catalytic performance for the designed catalyst in comparison to the benchmark catalyst. Highly uniform distribution of active sites in the molecular organic framework in addition to the presence of required CO_2 and H_2 sorption sites were found to be the key reasons for its high catalytic activity, selectivity and stability of the reported system in this study.

Acknowledgements

The authors gratefully acknowledge the financial support, resources and facilities provided by the King Abdullah University of Science and Technology (KAUST).

SPRING22
+18TH GCPS
A Joint AIChE and CCPS Meeting

Carbon Capture Vs Lithium-Ion Batteries – Which Pairs Better with Renewable Electricity?

Michael C. Stern
Exponent, Inc.
1075 Worcester Street, Natick, MA 01760
mstern@exponent.com

Michael Orella,[a] Achim Wechsung,[a] Christopher Buehler[b]
Exponent, Inc.
a – Natick, Massachusetts
b – New York, NY

Prepared for Presentation at
American Institute of Chemical Engineers
2022 Spring Meeting and 18th Global Congress on Process Safety
San Antonio, TX
April 10 – April 14, 2022

Z210111164847 - 9205

Carbon Capture Vs Lithium Ion Batteries – Which Pairs Better with Renewable Electricity?

Michael C. Stern
Exponent, Inc.
1075 Worcester Street, Natick, MA 01760
mstern@exponent.com

Michael Orella, Achim Wechsung, Christopher Buehler
Exponent, Inc.

Keywords: Carbon Capture & Storage, Power Systems and Sustainability Metrics for Energy Systems

Abstract

Utilities are under increased pressure to decarbonize their electricity production. As a result, utilities are increasing the share of intermittent renewable power sources including wind and solar. Meanwhile, many states and countries are reducing their dependence on nuclear power. These trends are creating an environment where utilities are expected to provide reliable power based on increasingly intermittent sources. To achieve the desired level of reliability, utilities will need technologies that can bridge the differences between when power is being produced versus when power is being consumed. Grid scale lithium-ion batteries have received significant attention for this purpose while others suggest that fossil-fuel based systems with carbon capture could provide greater reliability in the event of prolonged disruptions due to extreme weather.

1 Intermittent Renewable Sources

Three primary types of intermittent renewable sources are solar, wind, and hydroelectric power. Other renewable sources that may be less intermittent include geothermal and biomass (depending on biomass availability).

Solar power fluctuates on several timescales.[1] First, solar energy varies in a predictable way due to the diurnal cycle each day. Secondly, weather patterns on the order of minutes to days may cause rapid and/or prolonged drops in solar availability. Coarsely, these weather-based changes may be predictable days in advance but are challenging to accurately predict weeks or months in advance. In cooler climates where ice or snowstorms can occur, these can lead to day- to week-long outages or reductions. Solar thermal plants, based on the thermal mass of their systems, are more resilient to short-term variations. Third, solar availability varies seasonally (i.e., on the order of months) due to predictable changes in the amount of daylight and the incident solar irradiance.

Like solar, wind energy availability also varies predictably as a result of diurnal and seasonal cycles as well as less predictably due to weather-based changes.[2] Wind energy also, however, is subject to even shorter-term fluctuations because of wind gusts.

Hydroelectric power is not subject to diurnal cycles or short-term weather impacts but may be subject to both long-term weather patterns (i.e., droughts) and predictable seasonal changes in water levels.[3] Water releases may also need to be coordinated with ecological considerations around migrating fish such as salmon. As water levels drop due to drought or other factors, the amount of energy that can be generated by allowing a given amount of water to pass through a hydroelectric turbine decreases. Further, when water elevations reach critical levels, the amount of water allowed to pass the dam may be restricted to protect the availability of water for drinking or other purposes.

While each of the above sources has significant variability on certain timescales, much of the variability can be eliminated by aggregation of multiple generation locations at distant locations.[4] The longer the timescale of the variability, the farther apart the locations must be to effectively smooth the power generation curve. Due to the practical limits of power transmission not all variations can be smoothed. Diurnal and seasonal variation are particularly resistant to smoothing. Weather patterns, which often move relatively slowly compared to the speed of diurnal cycle, can generally be largely mitigated using multiple distant generation locations.

2 Existing Grid-Scale Energy Storage

Energy storage technologies have been available for decades with the primary technology being pumped storage hydropower (PSH).[5] The current US PSH fleet includes the ability to provide or absorb 22 GW of power with a storage capacity of 550 GWh. This represents approximately 95% of all utility-scale storage in the United States.

PSH efficiency can exceed 80% but can be capitally intensive to construct. PSH has relatively low energy density compared to chemical technologies (like batteries or hydrogen) at around 1 watt-hour per liter.[6] Other technologies including compressed air energy storage (CAES) and flywheels have been used in limited applications.

While PSH is likely to continue to play a critical role in future energy storage applications, the relatively long construction times, geographical constraints, and low energy density of PSH will limit its adoption. As a result, technologies such as battery storage and excess thermal capacity fitted with carbon capture and sequestration (CCS) are being considered for integration of high quantities of renewable power generation with a low carbon intensity.

3 Carbon Capture for Grid Stability

CCS on combustion-based power systems (coal, gas, oil, biomass, biogas) provide a means for stable power generation with significantly reduced emissions, and in some cases with bio-derived fuels, even negative net emissions.[7] CCS of the CO_2, however, is not without significant challenges. The primary challenges are the significant parasitic power loads of operating the carbon capture plant, the large capital cost, the difficulty of retrofitting existing facilities, and the need for proximate CO_2 transport and storage infrastructure. There are a number of pilot-scale technologies currently being evaluated for their commercial viability, but recent history also includes several failures of large-scale CCS projects due to substantial cost overruns during construction and operation.[8]

CCS systems, especially those based on natural gas-fueled power plants, can adjust to the demands of the grid when power from intermittent sources fluctuates. In addition, CCS systems could adjust the extent of carbon capture to provide further flexibility during extreme surges in energy demand. For example, during low energy demand, the carbon capture rate could be increased to above 90% and during infrequent extreme or emergency demand situations, the carbon capture rate could be reduced to increase the net power output of the facility.

Power plants with CCS facilities can rely on pre-existing infrastructure for the extraction, transport, and local storage of fossil fuels for their energy supply. The reliability of these systems is largely decoupled from the factors that affect the supply of renewable power sources making them an attractive pairing for reliability. Even during prolonged periods of low renewable energy generation, the fossil fuel systems should be able to continue to support the grid.

Many CCS systems for power generation are designed for around 90% capture. However, the reduction in CO_2 intensity is less than 90% because the parasitic energy consumption of the CCS system increases the total fuel required.[9] In the case of natural gas fired plants, capture of the post-combustion carbon cannot address methane emissions from the extraction and transport of natural gas. Depending on the source of the natural gas, these upstream emissions may be a significant factor in the overall footprint of the power generation. Further, CCS systems may require constant replenishment of sorbent chemicals that have associated carbon footprints. A CCS system may also have other environmental and health impacts including increased water usage and the release of small amounts of chemicals such as nitrosamines.

4 Lithium-Ion Batteries for Grid-Scale Energy Storage

Lithium-ion (Li-ion) batteries have garnered increasing interest as a potential low-cost method for storing chemical energy at the grid scale. Estimates of the round-trip efficiency for utility-scale Li-ion battery storage facilities are comparable to PSH, near 80%.[10] However, unlike PSH, Li-ion batteries are not geographically constrained, and boast energy densities greater than 250 watt-hours per liter.[11]

Li-ion battery systems may be used to respond to intermittent renewable power sources, such as solar and wind, as these grow in the share of electrical power-generating capacity. As discussed above, wind and solar power-sources may display diurnal variation, which Li-ion batteries are well-suited to handle, due to a typical charge/discharge cycle for these batteries occurring over the course of several hours.

However, an energy storage system that relies solely on Li-ion batteries may need to be substantially over-designed to handle extreme deviations from the average behavior (e.g., cases of prolonged cloud cover). In addition, as extreme weather events fueled by climate change occur with greater frequency, the probability of a need for such over-designed systems to handle their limiting cases increases. Lastly, long-duration storage that would be needed for seasonal energy storage remains a significant challenge for Li-ion batteries.[12] Li-ion batteries appear well-poised to play an important role in energy storage systems for increasingly intermittent electrical power sources.

5 Hydrogen

Electricity can be converted into hydrogen by means of electrolysis. Hydrogen can be converted back into electricity using either fuel cells or combined-cycle gas turbines. In principle, hydrogen can be stored at relatively low cost for long periods of time. Estimates of the electrolysis system efficiency are on the order of 75%.[13] Large combined-cycle natural gas turbines operate at efficiencies close to, or just exceeding, 60%, while the first hydrogen-fueled demonstration plants currently achieve 42%.[14,15] Fuel cells can achieve similar electrical efficiency.[16] Overall, the combined efficiency is approximately 45%.

One conceivable application for such a system is electrical grid balancing similar to pump-storage hydro power plants. In order to justify the capital expenditure and the inherent efficiency losses, these plants provide power to the electrical grid at times of high demand and consequently high prices whereas they produce hydrogen at times of low demand and lower prices.

However, it may be even more advantageous to not convert hydrogen back to electricity, but rather use it directly. Then, the electrolysis plant operates as a "peak supply shaving" plant. Rather than operating at times of high demand, it takes up power from intermittent sources when these generate more power than demanded. Hydrogen can be stored, e.g., in underground caverns, and fed into existing hydrogen pipeline networks over time.

6 Conclusions

The effective and reliable deployment of high fractions of renewable energy will require pairing with other sources to convert intermittent sources to meet the continuous demand of the power grid. No single technology is likely to be sufficient due to the widely disparate timescales that must be considered. While batteries seem well-positioned for daily variability, other technologies would be required to account for longer term variability such as seasonal variations in solar and wind output or prolonged weather systems that may be exacerbated by climate change. Currently, gas peaker plants manage these disruptions and may be difficult to retrofit with CCS due to the capital intensity of CCS technologies.

7 References

[1] Shedd S, Hodge B-M, Florita A, Orwig K. A Statistical Characterization of Solar Photovoltaic Power Variability at Small Timescales. NREL CP-5500-56165, 2012.

[2] Yan YH. Long-Term Wind Power Variability. NREL TP-5500-53637, 2012.

[3] Tarroja B, Forrest K, Chiang F, AghaKouchak A, Samuelsen S. Implications of hydropower variability from climate change for a future, highly-renewable electric grid in California, *Applied Energy* 2019; **237**.

[4] Engeland K, Borga M, Creutin J-D, François B, Ramos M-H, Vidal J-P. Space-time variability of climate variables and intermittent renewable electricity production – A review, *Renewable and Sustainable Energy Reviews* 2017; **79**.

[5] US Department of Energy, Water Power Technologies Office. How Pumped Storage Hydropower Works. https://www.energy.gov/eere/water/how-pumped-storage-hydropower-works. Accessed April 2, 2022.

[6] Environmental and Energy Study Institute. Fact Sheet | Energy Storage (2019). https://www.eesi.org/papers/view/energy-storage-2019#:~:text=Pumped%2Dstorage%20hydropower%20is%20more,hours%20for%20lithium%2Dion%20batteries. Accessed April 2, 2022.

[7] Heuberger CF, Staffell I, Shah N, Mac Dowell N. A systems approach to quantifying the value of power generation and energy storage technologies in future electricity networks. *Computers & Chemical Engineering* 2017; **107**.

[8] Abdulla A, Hanna R, Schell KR, Babacan O, Victor DG. Explaining successful and failed investments in U.S. carbon capture and storage using empirical and expert assessments. *Environ Res Lett* 2020; **16**: 014036.

[9] Systems Analysis and Technology Evaluation (IEF-STE). Environmental Evaluations of CCS using life cycle assessment (LCA). IEAGHG 2010/TR2, May 2010. https://ieaghg.org/docs/General_Docs/Reports/2010-TR2.pdf. Accessed April 2, 2022.

[10] Mey A. Utility-scale batteries and pumped storage return about 80% of the electricity they store. US Energy Information Agency. February 12, 2021. https://www.eia.gov/todayinenergy/detail.php?id=46756. Accessed April 2, 2022.

[11] University of Washington Clean Energy Institute. What is a lithium-ion battery and how does it work?. 2020. https://www.cei.washington.edu/education/science-of-solar/battery-technology/. Accessed April 2, 2022.

[12] Serpell O. The opportunities and limitations of seasonal energy storage. Kleinman Center for Energy Policy, Nov 2020. https://kleinmanenergy.upenn.edu/wp-content/uploads/2020/10/KCEP-Opportunities-and-Limitations-P10.pdf. Accessed April 2, 2022.

[13] US Department of Energy, Hydrogen and Fuel Cell Technologies Office. DOE Technical Targets for Hydrogen Production from Electrolysis. https://www.energy.gov/eere/fuelcells/doe-technical-targets-hydrogen-production-electrolysis. Accessed April 2, 2022.

[14] Shiozaki S, Fujii T, Takenaga K, Ozawa M, Yamada A. Gas turbine combined cycle. *Advances in Power Boilers* 2021; 305-344.

[15] Fusina: Achieving low NOx from hydrogen combined-cycle power. *Power Engineering International* October 1, 2010. https://www.powerengineeringint.com/world-

regions/europe/fusina-achieving-low-nox-from-hydrogen-combined-cycle-power/. Accessed April 2, 2022.

[16] US Department of Energy, Hydrogen and Fuel Cell Technologies Office. Comparison of Fuel Cell Technologies. https://www.energy.gov/eere/fuelcells/comparison-fuel-cell-technologies. Accessed April 2, 2022.

SPRING22
+18TH GCPS
A Joint AIChE and CCPS Meeting

Quantifying Atmospheric Oxygen Depletion and the Increase of Global Temperature Caused by Natural Gas, Hydrogen and Nuclear Power

Dr Frank J. Elvin
Texas Specialty Catalysts
Bellaire Texas
elvini@flash.net

Prepared for Presentation at
American Institute of Chemical Engineers
2022 Spring Meeting and 18th Global Congress on Process Safety
San Antonio, TX
April 10 – April 14, 2022

Quantifying Atmospheric Oxygen Depletion and the Increase of Global Temperature Caused by Natural Gas, Hydrogen and Nuclear Power

Dr Frank J. Elvin
Texas Specialty Catalysts
Bellaire Texas
elvini@flash.net

Keywords: Sustainability, Global Warming, Atmospheric Oxygen Depletion

Abstract

This paper analyzes the rate at which natural gas, hydrogen, nuclear fission and nuclear fusion can be used before they contribute significantly to oxygen depletion and global warming. It discusses how exothermic and endothermic reactions will affect global radiation and temperatures after greenhouse gas concentrations have been reduced to their pre-industrial levels.

The hydrogen in fossil fuels reacts with oxygen, removing oxygen from the atmosphere while releasing 286KJ/mol of heat. The carbon in fossil fuels reacts with oxygen to produce carbon dioxide, releasing 394KJ/mol of heat. Currently half of this carbon dioxide reacts with water in plants to produce oxygen and sugars. The energy absorbed in this photosynthetic process is 467KJ/mol of carbon dioxide. The oxygen released by photosynthesis equals the oxygen consumed during carbon oxidation. Therefore, the fossil fuel carbon that is converted into carbon dioxide and subsequently converted into glucose reduces atmospheric heat by 73KJ/mol and causes no oxygen depletion. The remaining carbon dioxide produced from fossil fuels accumulates in the atmosphere depleting oxygen and adding reaction heat. Nuclear power doesn't cause oxygen depletion but does add heat to the atmosphere.

1 Introduction

Prior to the use of coal for heating, wood was the primary source of fuel. Coal did not replace wood until the 1700s. The consumption of coal rapidly increased in the 1800s with the demands of the industrial revolution initiating global warming and atmospheric oxygen depletion. Unlike burning coal burning wood was balanced by photosynthesis, which removed carbon dioxide, replenished the oxygen and removed the heat of combustion. It has taken 300 years for CO2 to cause noticeable global warming as shown in figures 1 and 2. The relationship between the mean global surface temperature anomaly (TA)[1] and atmospheric CO2 concentration (CO2)[2] is linear over the range of CO2 concentrations experienced on Earth: TA=0.01035*CO2-3.324. This linear equation was used to extrapolate the effects of changing atmospheric CO2 concentrations. Based on the Energy Information Agency (EIA) projected fossil fuel emissions data[3] atmospheric CO2 concentration will be 482ppm in 2050[4].

Figure 1 Earth's Surface Temperature Anomaly vs Year

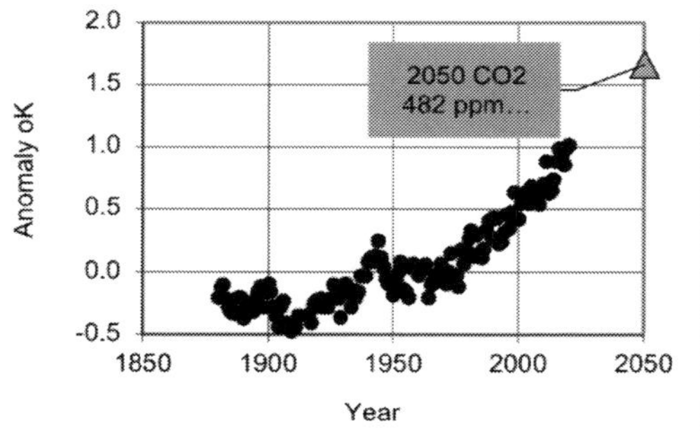

Figure 2 Earth's Temperature Anomaly vs CO2 ppm

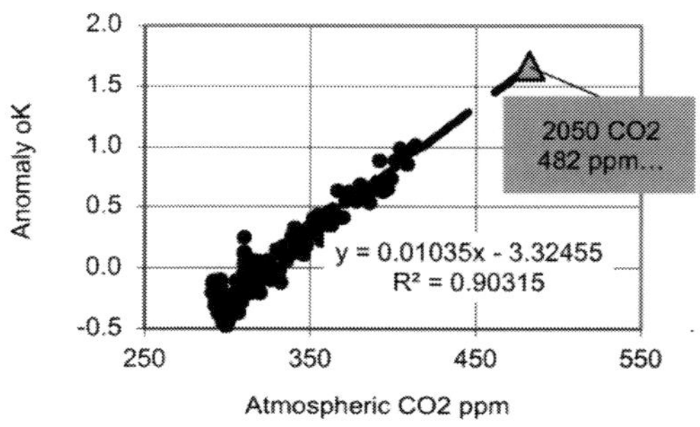

If CO2 and other greenhouse gas (GHG) emissions are reduced to pre-industrial levels any global warming caused by GHGs can be reversed. However, if GHGs are reduced by using nonrenewable sources such as hydrogen from fossil fuels (blue hydrogen), nuclear fission or fusion, or by using fossil fuels with carbon capture and sequestration (CCS), global warming and oxygen depletion will occur sometime in the future. The objective of this paper is to estimate when this could occur and how severe it could be.

Atmospheric CO2 concentrations are predicted using an equation derived from published data for 1750 to 1980[5]. The equation is ΔCO2=5.13+F-0.0177*CO2, where CO2 is the predicted atmospheric CO2 concentration for January of a given year, ΔCO2 is the predicted ppm change in CO2 in that year, and F is the ppm of fossil fuel CO2 emissions in that year (Gt/7.9). Every value for CO2 was predicted by performing sequential annual predictions of ΔCO2 and CO2 using the equation CO2(year(n+1)=CO2(year(n)+ΔCO2(year(n)). Atmospheric CO2 concentrations were predicted for 1751 to 2021 using only the 1750 ice core CO2 measurement of 278 ppm[6] and published annual global CO2 emissions data for 1750 to 2021[7]. Measured and predicted values of CO2 are shown in figure 3. Because of the accuracy of CO2 predictions, the two equations described above are used to estimate caubon related oxygen depletion.

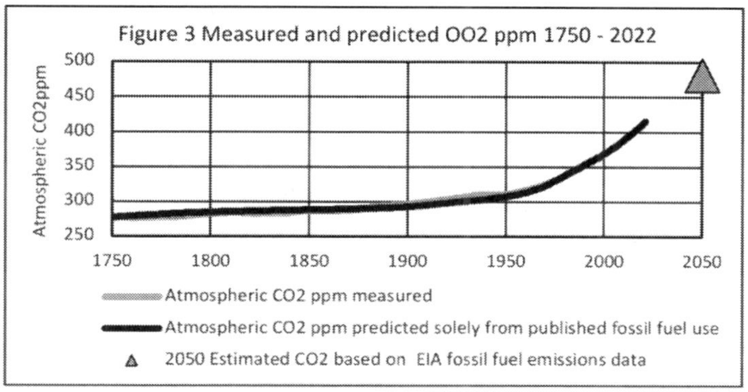

Only a few of the many possible future energy scenarios are described in this paper. The base 2021 global fuel mix and energy consumption was taken from a BP publication(8). The global CO2 emissions calculated from BP data was reconciled with EIA's estimated global CO2 emissions(3). Recent global energy consumption has increased by 1.8% per year. However, future increases will be different, and the 1.8% and 1% annual global energy increases used throughout this study are not predictions. Some of the global fuel mix scenarios, such as 100% coal, 100% hydrogen, and 100% CCS are also for information and comparison only. This paper shows how different fuels, even carbon neutral fuels affect the rate of global warming and oxygen depletion. The spreadsheets for calculations of atmospheric CO2 concentrations, temperatures and atmospheric oxygen levels are available on request.

2 Mean Global Temperature

The mean global surface temperature T_1 in °K is calculated from the equation $R=A.e.\sigma.T_1^4$ where A is the Earth surface area of $5.10*10^{14}$ m^2, e is the current effective emissivity of 0.59, σ is the Stefan Boltzmann constant of $5.67*10^{-8}$. R is assumed to be constant at 230.1 W/m^2 (for a total thermal energy radiating into space of $1.17*10^{17}$ W or $3.7*10^{18}$ MJ/year). If the CO2 concentration in the atmosphere remains constant e and T will remain constant. If the atmospheric CO2 concentration continues to increase emissivity will decrease and T will increase to balance solar radiation. The low overall emissivity of 0.59 is due to water vapor, CO2, other GHGs, clouds, dust and aerosols. The reduction in emissivity from 0.61to 0.59 is assumed to be caused by the increase in atmospheric CO2 concentration since 1750(12).

The mean global atmospheric temperature T_2°K at the Earth's surface, referred to simply as "atmospheric temperature" in this paper, is the mean global surface temperature T_1 plus the differential temperature ΔT between the Earth's surface and the atmosphere in contact with the Earth's surface. ΔT is the heat transfer driving force for energy added to the atmosphere from global energy consumption that is ultimately radiated through the atmosphere to space. When there is no heat transfer between the atmosphere and the Earth's surface ΔT is assumed to be zero.

Heat transfer rates within the oceans and land were calculated and are available on request. Calculations show that they did not significantly affect radiation and have not been included in this paper. Any effects of differences between ocean and land temperatures on radiation were calculated to be insignificant and are not included.

The global energy consumption from all sources is currently 18.1 Terawatt (TW)[9] ($5.708*10^{14}$ MJ/year). This energy is dispersed in the atmosphere, raising atmospheric temperature until the differential temperature between the atmosphere and the Earth's surface ΔT is sufficient to transfer the heat to the Earth's surface. The equation $\Delta T=H/(U.A)$ calculates this differential temperature, where H is the annual heat from all global nonrenewable energy sources, A is Earth's surface area and U is the overall heat transfer coefficient of $5W/(m^2.^oK)$ or $157.68Mj/(year.m^2.K)$. The atmospheric temperature increases are all relative to 288°K, the mean global surface and atmospheric temperature in 2021. The increase in atmospheric temperature equals the sum of the increases of mean global surface temperature caused by GHGs and global energy plus the differential temperature between atmosphere and Earth's surface.

When GHGs are the sole cause of global warming, the heat radiation into space equals net solar input, which is assumed constant. This changes when heat from nonrenewable energy is sufficient to heat up the atmosphere. This heat is transferred to the Earth's surface and radiated through the atmosphere into space. The potential atmospheric temperature increase for global power consumption is shown in Figure 4. As the current global energy consumption is 18.1 TW, it will be several hundred years before global energy use reaches 2,000 TW and affects atmospheric temperatures. If 2,000 TW of energy consumption was achieved using carbon or hydrogen-based fuels, then oxygen depletion rather than temperature will limit further energy growth. Figures 5 to 11 show estimated atmospheric temperatures for different annual energy increases, different fuel combinations and show the effect of 100%CCS. Some of these estimates are infeasible, especially those for 1.8% annual global energy increases, which could be limited by oxygen depletion by year 2300.

Figure 4 Atmospheric Temperature increase for differnt Terawatts (TW) of nonrenewable energy when used with 100% CCS

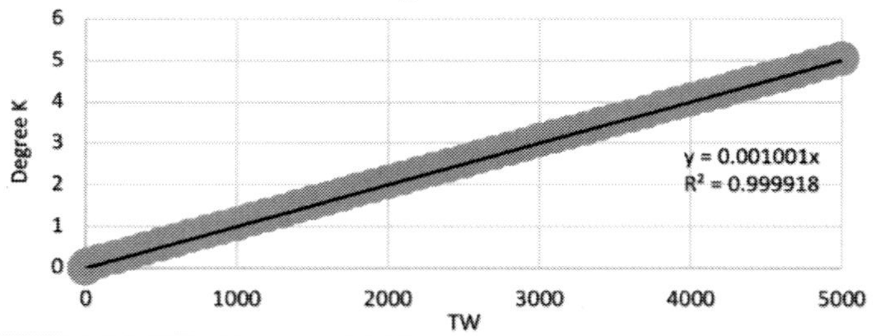

2.1: Base case: Fuel mix percentages constant (figure 5)

The fuel mix used in figure 5 is BP's 2021 estimate. The graph shows how the atmospheric temperature would change if the mix proportions remained constant through 2100. If this were to happen, the temperature would increase rapidly whether the total global energy use increased 1% or 1.8% per year without CCS. With 100% carbon sequestration the temperature increases reflect the effect of nonrenewable energy on global temperature, as there is no CO2 effect. At the 1.8% annual energy increase it would take 200 years before there was a significant temperature increase and at the 1% annual increase beyond 400 years. The fuel mix for case 2.1 is: wind, solar and biomass 4.3%; hydro 6.2%, nuclear 3.8%; coal 27.9%; natural gas 24.1%; and crude oil 33.7%. Although 100% CCS is improbable, it is a best-case scenario, and all CCS cases will result in higher atmospheric temperatures than the 100% CCS case.

2.2 Fuel mix gradually changing to renewable energy (figure 6)

The fuel mix change from 2021 to 2050 used for this case is: +10%/year increase in wind, solar and biomass; hydro and nuclear power constant; coal +1%/year: natural gas +3%/year and Crude Oil by difference (an average of -2%/year). After 2050 hydro and nuclear remain constant: coal, natural gas and crude oil -3%/year. Coal, crude oil and natural gas are not eliminated totally. A strategic production of between 0.3-1Gt/year for each fuel is maintained. Note that with 100% CCS there is insufficient nonrenewable energy to cause the temperature to increase and the the1% and 1.8% plots (the blue and grey data points) coincide.

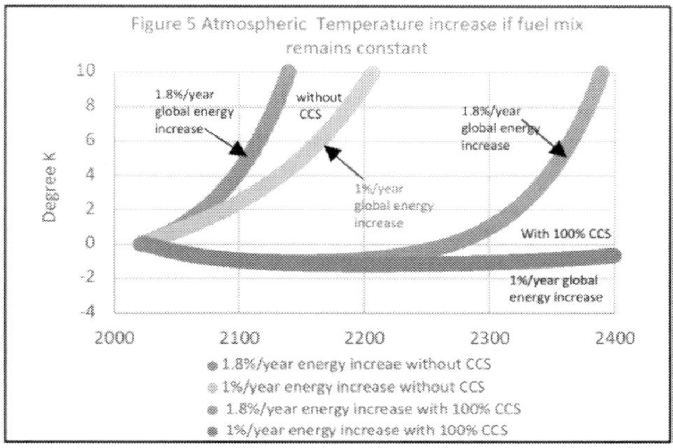

Figure 5 Atmospheric Temperature increase if fuel mix remains constant

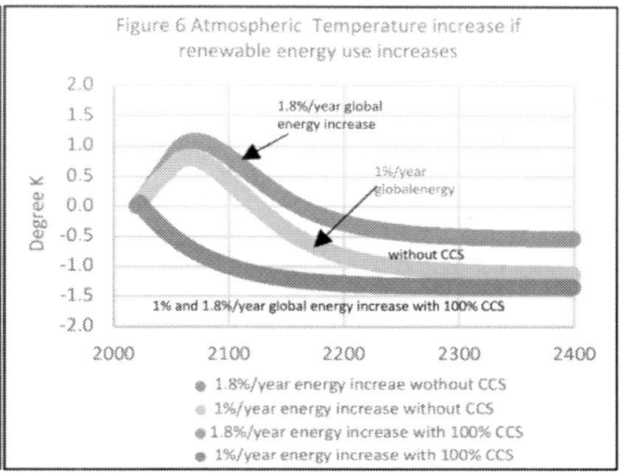

Figure 6 Atmospheric Temperature increase if renewable energy use increases

2.3 Changing to 100% coal after 2021 (figure 7)

Although the use of 100% coal is infeasible, this case has been presented as a hypothetical worst-case scenario for comparison with the ***2.1: Base case***

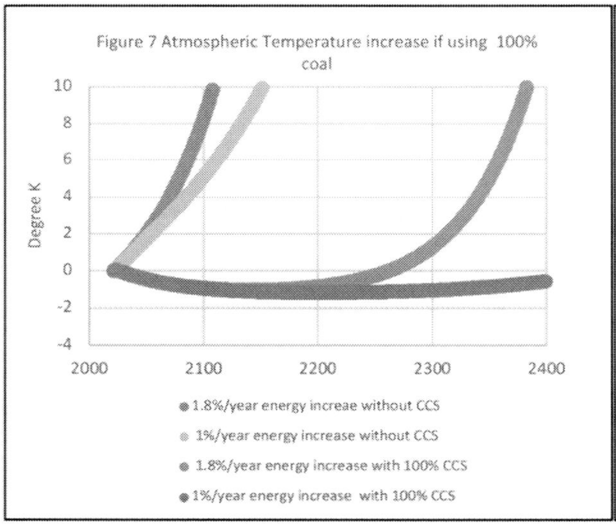

Figure 7 Atmospheric Temperature increase if using 100% coal

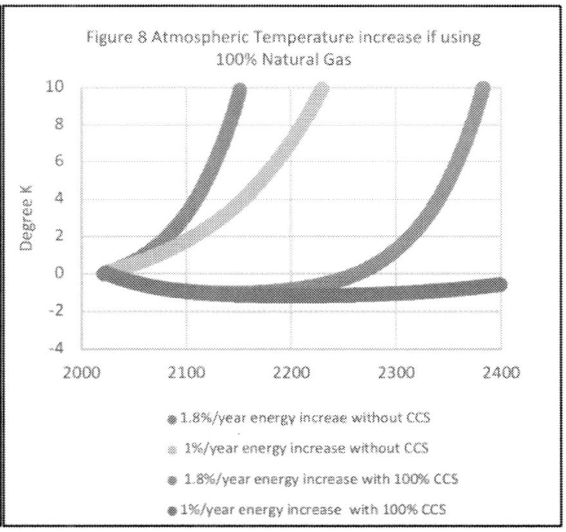

Figure 8 Atmospheric Temperature increase if using 100% Natural Gas

With 100% coal without CCS the atmospheric temperature increases 10°K 50 years earlier than with the current fuel mix ***(2.1 base case)*** without CCS, because coal has more CO2 per unit energy than the current fuel mix. However, with 100% coal and 100% CCS the atmospheric temperature increases10°K which is similar to the current fuel mix ***(2.1 base case Figure 5)***. This is because CCS removes the effect of CO2 on temperature and the nonrenewable energy input to the atmosphere are similar for these two cases.

2.4 *Changing to 100% Natural Gas after 2021(figure 8)*

Without CCS the temperature increases 10°K 50 years later with 100% natural gas than the 100% coal ***(case (2.3 Figure 7)*** because coal has more CO2 per unit energy than natural gas. The 100% natural gas case with 100% CCS is similar to 100% coal with 100% CCS because the CO2 effect is eliminated leaving only the energy effect Because the energy in the natural gas and the coal cases are nonrenewable the effect of energy on temperature is greater with 1.8% annual global energy increase than the 1% annual global increase.

2.5 Changing to 100% Blue Hydrogen after 2021 (figure 9)

Blue hydrogen produced from natural gas assumes that the CO2 produced in its manufacture undergoes CCS. However, unlike green hydrogen produced by electrolysis from renewable power such as wind or solar, blue hydrogen energy adds heat to the atmosphere as shown in figure 9

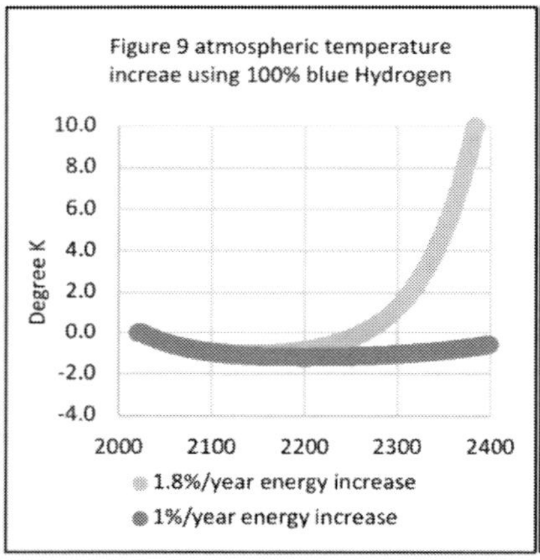

Figure 9 atmospheric temperature increae using 100% blue Hydrogen

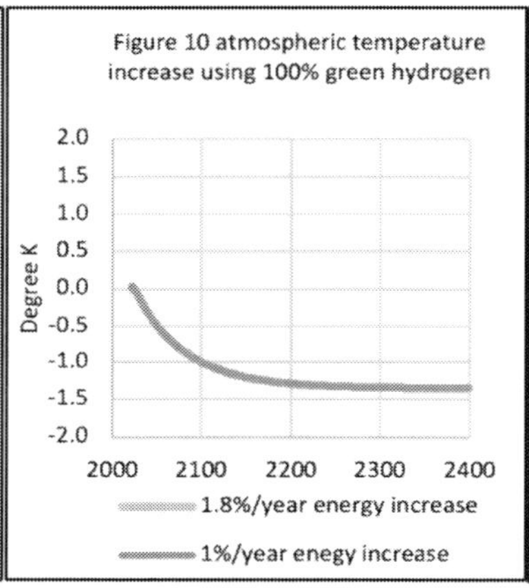

Figure 10 atmospheric temperature increase using 100% green hydrogen

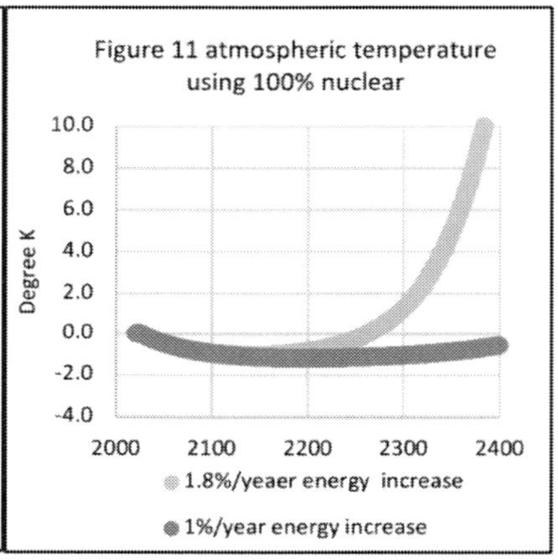

Figure 11 atmospheric temperature using 100% nuclear

2.6 Changing to 100% Green Hydrogen after 2021(figure 10)

Changing to green hydrogen eliminates all CO2 and CH4 and the global warming resulting from these GHGs. Atmospheric CO2 would decrease to pre-industrial levels as shown in figure 12. The atmospheric temperature and CO2 concentration is independent of energy consumption, making green hydrogen the perfect fuel if global energy consumption is insatiable.

2.7 Changing to 100% Nuclear after 2021 (figure 11)

Nuclear power eliminates the CO2 greenhouse effect but would need to be limited to 2000 TW to prevent another 2°K of global warming. This would take till 2300 with a 1.8%/year annual energy increase. This would be more than one hundred times current energy consumption, which is unlikely if not impossible to achieve with nuclear fission of fusion.

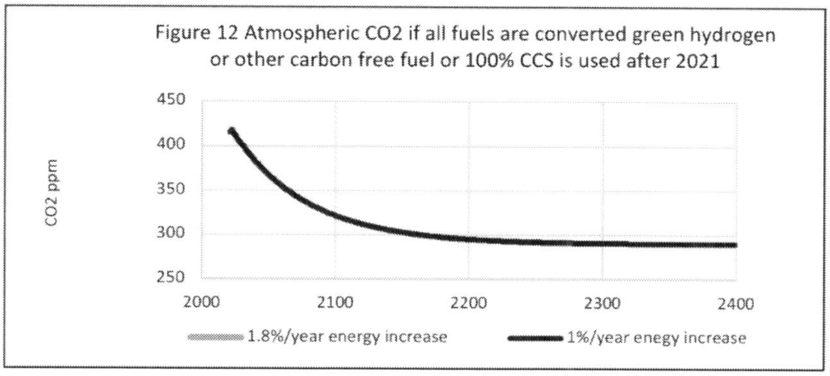

Figure 12 Atmospheric CO2 if all fuels are converted green hydrogen or other carbon free fuel or 100% CCS is used after 2021

3 Atmospheric Oxygen

All cases presented assume a 1% increase in annual global energy use. Carbon converted to CO2 but not subsequently converted to oxygen by photosynthesis depletes atmospheric oxygen. Hydrogen converted to water but not subsequently converted back to hydrogen and oxygen also depletes atmospheric oxygen. Fossil fuels contain both carbon and hydrogen. Oxygen depletion caused by carbon oxidation is calculated from the measured annual CO2 increase in the atmosphere and estimated carbonate increase in coral, shells and bones. CO2 not accumulating in the atmosphere or being converted to $CO3^{=}$ in the oceans is assumed to be converted to sugars, cellulose and oxygen by photosynthesis. Oxygen depletion caused by hydrogen is calculated directly from the hydrogen in fossil fuels and any blue hydrogen used as a fuel. Green hydrogen produced by photosynthesis from renewable fuels is considered oxygen neutral. To calculate the oxygen removal by hydrogen, the following molecular formulae have been assumed: coal CH; natural gas CH4; and crude oil CH2

3.1: Base Case, Fuel mix percentages constant after 2021 (figure 13)

With current fuel mix and a 1% annual increase in global energy it would take 200 years before there is a significant reduction in atmospheric oxygen. Oxygen depletion is accelerated by CCS because as atmospheric CO2 decreases the rate of photosynthetic oxygen production decreases.

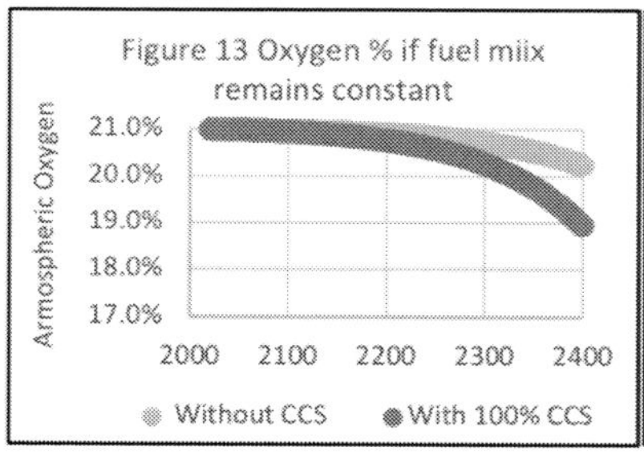

Figure 13 Oxygen % if fuel miix remains constant

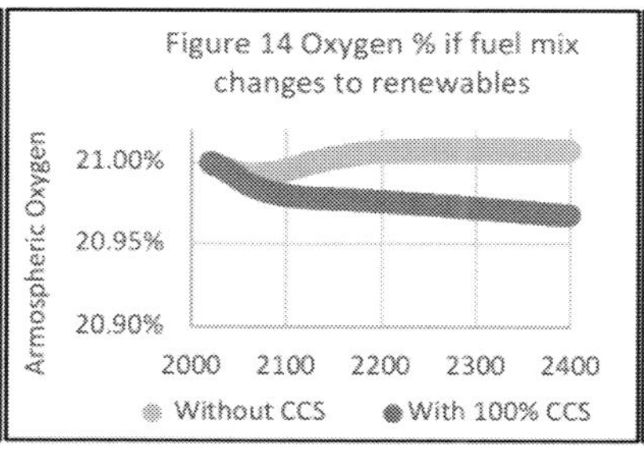

Figure 14 Oxygen % if fuel mix changes to renewables

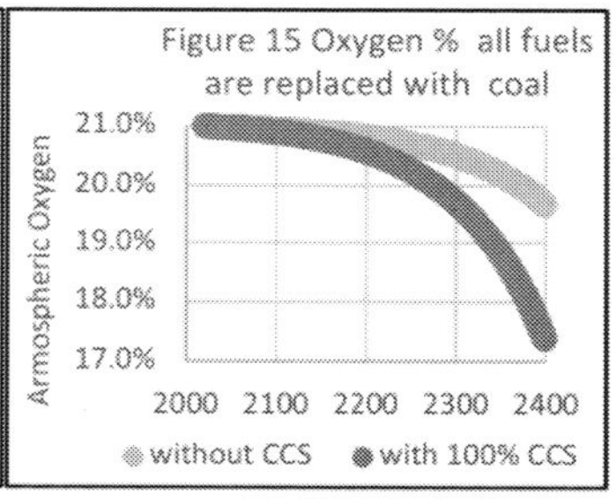

Figure 15 Oxygen % all fuels are replaced with coal

3.2: Fuel mix percentage gradually changing to renewables after 2021 (figure 14)

There will be little oxygen depletion if there is a gradual change to renewable energy sources.

3.3 Fuel mix percentage changing to 100% coal after 2021 (figure 15)

Burning coal adds more CO2 into the atmosphere than can be removed by photosynthesis increasing oxygen depletion. CCS increases oxygen depletion as it does not prevent the removal of oxygen during combustion but decreases the CO2 in the atmosphere, which decreases photosynthesis.

3.4 Fuel mix percentage changing to 100% Natural Gas after 2021(figure 16)

Natural gas depletes oxygen more than coal or the current fuel mix because of its higher hydrogen content. CCS of natural gas reduces the atmospheric CO2 concentration which reduces photosynthesis.

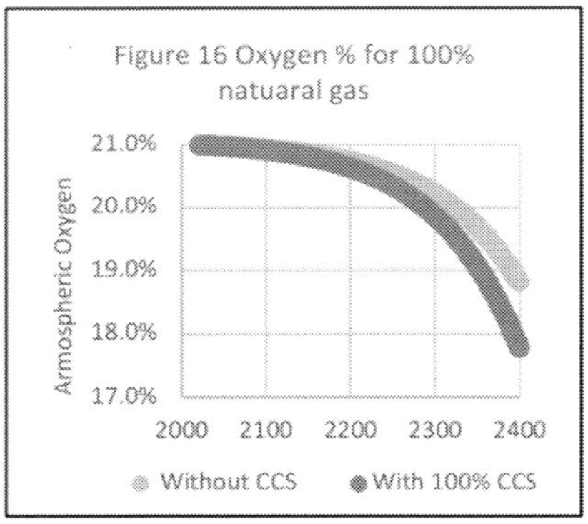

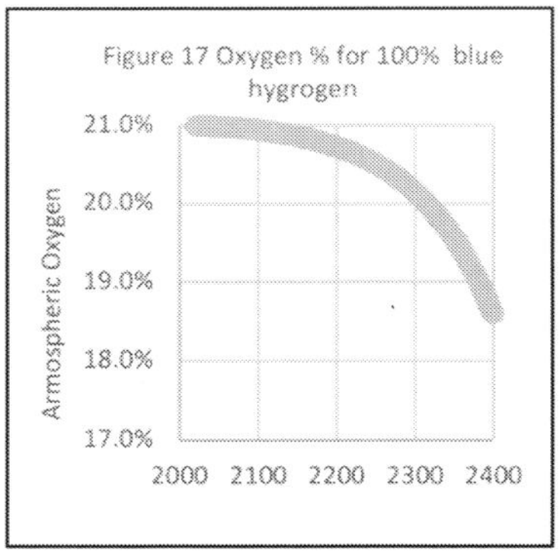

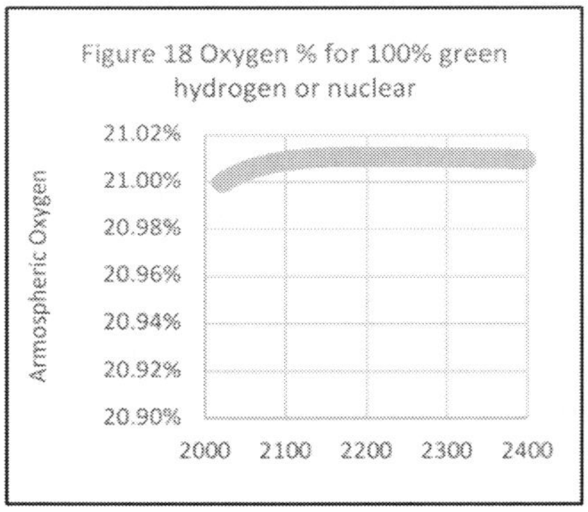

3.5 Fuel mix percentage changing to 100% Blue Hydrogen after 2021 (figure 17)

Blue hydrogen is produced from fossil fuels when the biproduct CO2 is captured and sequestrated. Blue hydrogen depletes atmospheric oxygen as much as natural gas. This is because none of the water formed by combustion is converted back to oxygen.

3.6 Fuel mix percentage changing to 100% Green Hydrogen, after 2021 (figure 18)

Green hydrogen does not deplete atmospheric oxygen if all the biproduct oxygen produced by electrolysis of water is returned to the atmosphere.

3.7 Fuel mix percentage changing to 100% Nuclear after 2021 (figure 18)

Nuclear power (fission or fusion) does not deplete atmospheric oxygen.

3.8 Atmospheric Oxygen and accumulated energy use from blue hydrogen (figure 19)

Whereas atmospheric temperature is a function of nonrenewable energy rate as shown in figure 4, atmospheric oxygen depletion is a function of the accumulated consumption of carbon and hydrogen-based fuels and not their rate of consumption. It does not matter how slowly blue hydrogen or natural gas are used, eventually they will consume all of Earth's available atmospheric oxygen.

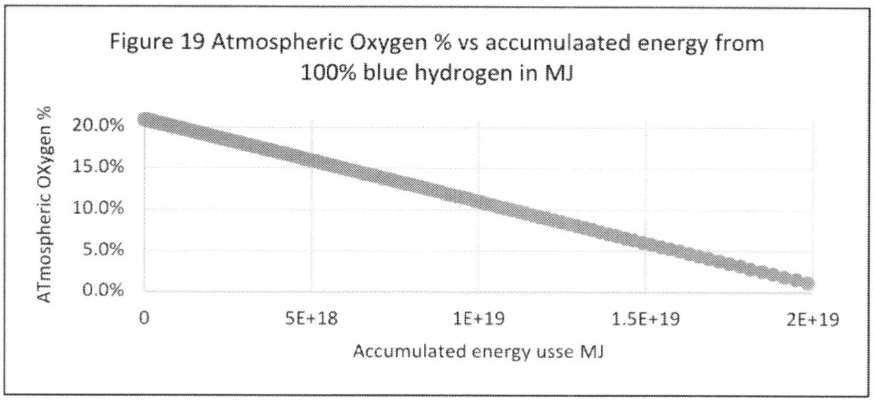

4 Conclusions

The reduction of Earth's thermal radiation emissivity caused by CO2 emissions is currently the primary cause of annual global warming. The rate of CO2 emissions can be reduced by reducing global energy use, reducing the use of carbon-based fossil fuels and the application of CCS. However, some methods used to reduce atmospheric CO2 concentrations, such as natural gas with CCS and blue hydrogen, can cause global warming and atmospheric oxygen depletion that are unrelated to GHG emissions. The rate of nonrenewable energy consumption must be kept below 2,000 TW to prevent energy related global warming. The accumulated use of energy from natural gas and blue hydrogen must be less than $4.5*10^{18}$ MJ. Global warming and oxygen depletion can only be prevented by the elimination of nonrenewable energy sources.

5 References

1. National Centers for Environmental Information NOAA FAQ February 202
2. National Oceanic and Atmospheric Administration (NOAA) Earth System Research Laboratory ESRL atmospheric CO_2 data published July 16 2012. Data added for 2012-2022
3. Energy Information Administration (EIA) International Reports 2006 -2018). International Energy outlook 201. 3 July 35 2013. Report Number: DOE/EIA-0484(2013) International Energy Outlook 2019 David Daniels EIA presentation Rice university October 2 2019
4. Atmospheric CO2 Predictions for Different CCS Scenarios Based on EIA Fossil Fuel Emissions Projections F, J, Elvin AIChE Spring Meeting 2020
5. Effects of Fossil Fuel Emissions and Seawater Absorption on Atmospheric Carbon Dioxide Concentrations F. J. Elvin AICHE Spring meeting 2018 Orlando Florida.
6. Carbon Dioxide Information Analysis Center (CDIAC) Historical CO_2 record from the Law Dome DE08, DE08-2, and DSS ice cores June 1998 D.M. Etheridge
7. Carbon Dioxide Information Analysis Center (CDIAC): Global CO_2 Emissions from Fossil-Fuel Burning, Cement Manufacture, and Gas Flaring: 1751-2010 July 30, 2013 Tom Boden, Carbon Dioxide Information Analysis Center Oak Ridge National Laboratory
8. BP Energy Outlook 2020
9. BP Statistical Review of World Energy 2021

640206 Hydrogen Plasma Reduction of REOs/Salt for Rems Production

Abstract:

Rare earth elements (REEs) are crucial to the production of green energy motors in wind turbines, hybrid cars, electric vehicles, and electronic devices. REEs are also used in defense systems, including precision-guided munitions, radar systems, avionics, night vision equipment, satellites and guided missiles. China dominates the REEs production and exports and is continuing to reduce REEs exports. DoD and DOE are looking at options to increase domestic production in the U.S. However, rebuilding the U.S. rare earth industry could take many years.

In this paper, we will report a process for producing REMs from REOs by plasma-enhanced hydrogen reduction to recycle rare earth materials from post-consumer products. Of the five critical REEs, [dysprosium (Dy,66), neodymium (Nd,60), terbium (Tb,65), europium (Eu,63), and yttrium (Y,39)]. This is an efficient and low energy consumption technology for five most valuable REMs (Nd, Y, Dy, Sm, Ce) production from their corresponding oxides.

638743 Carbon Intensities and Criteria Air Pollutant Emissions of Waste- and Coproduct- Derived Bioenergy Resources Resulting from on-Road Transportation Applications

Abstract:

Biofuels are anticipated to represent an important portion of the alternative fuels for heavy-duty vehicles (HDVs) required for achieving the decarbonation goals in the United States. Increasing interest in waste- and coproduct-derived biodiesel, renewable diesel, and renewable natural gas (RNG) as alternative transportation fuels necessitates an understanding of carbon intensities and criteria air pollutants from the perspective of full supply chains. This study evaluates carbon intensities and criteria air pollutant emissions of biomass-based diesel (BBD) and RNG produced by different feedstocks as part of a project commissioned by the New York League of Conservation Voters. The feedstocks for BBD pathways contain cooking oil (UCO), distillers corn oil (DCO), tallow, and forest residue. Landfill gas or biogas from anaerobic digestion of animal waste and wastewater sludge are studied to quantify the emissions of RNG. The final biofuel products of BBD and RNG are used to evaluate their potential waste- or coproduct-to-wheel emissions of greenhouse gases (GHG) and criteria air pollutants related to feedstock recovery and transportation, fuel production, transportation and distribution, and use in HDVs. The system boundary is in accordance with the main life cycle stages covered by the fuel cycle module of the Greenhouse Gases, Regulated Emissions, and Energy Use in Technologies (GREET) Model by Argonne National Laboratory. The functional unit for comparison of different transportation fuels is 1 million British thermal unit (mmBtu) of heat delivered that allows for comparison across biofuels. Furthermore, after reviewing different studies and models including the GREET model and CA-GREET3.0, comparisons of the fuel cycle emission levels with different input and output assumptions are conducted to understand the wide range of the emission levels across different studies, allowing for understanding the hotspots of the biofuel pathways. Finally, the emissions of biofuels are compared against petrodiesel and fossil natural gas, allowing an understanding of potential reduction of the carbon intensities and criteria air pollutants as well as to make recommendations to further improve the emission levels of the studied biofuel pathways. While this study focuses on GHG emissions of BBD and RNG, the current findings also show that criteria air pollutant emissions of the biofuels are distinctively lower than their fossil-based counterparts (e.g., low sulfur diesel and liquefied natural gas). BBD has slightly higher emissions of nitrogen oxides (NOx) than petrodiesel; however, NOx can be controlled by a combination of selective catalytic reduction (SCR) with other exhaust after-treatments. The results will be presented at the conference.

639239 Green Hydrogen Production and Distribution System Simulation

Abstract:

The demand for today and the future is for true zero-emission power. Alternatives must be found to replace fossil fuels. Currently, batteries are a solution for automotive. Unfortunately, they are not suitable for many applications due to limitations with storage capacity, lifetime, charge constraints and environmental concerns. Therefore, green hydrogen production (produced for instance by electrolysis, using renewable electricity) is identified as a promising solution for long-term zero-emission renewable energy storage. Green hydrogen production simulation makes it possible to capture the complete process of green hydrogen production, predict interactions between subsystems and the global performances.

Green hydrogen production simulation can help address the challenges of green hydrogen production. An extensive multi-physics simulation platform is necessary that makes it possible to model complete systems. Sizing the different subsystems, considering various operating conditions is beneficial. It makes it possible to better integrate subsystems and improve the overall performances and ROI

639023 Modelling of Ammonia Decomposition Using Ru- and Co-Based Catalysts in a Packed Bed Reactor

Abstract:

Introduction

The carbon-free hydrogen economy is hinging on viable and feasible ways to transport hydrogen on-site. Currently, ammonia seems to be the most promising hydrogen carrier and is being explored through multiple routes in the literature; one of which being catalytic ammonia decomposition. Ammonia decomposition (AMD) has come up as an alternative for on-site, high purity hydrogen generation over the conventional reforming methods. Ru-based catalysts are proven to be the most active metal for the decomposition of ammonia, although costly. Recent works have also demonstrated at par conversions from Ba promoted Co-based catalysts against the Ru-based ones, with the former also showing higher stability over long reaction times. In this work, we develop an isothermal kinetic model for these two catalysts and explore the behavior of the decomposition process in a catalytic packed bed reactor in different operation regimes, with an aim of optimizing the reactor to maximize CO_x-free, high pressure hydrogen.

Materials and methods

The experimental data used for comparison and validation for both Ru-based & Co-based catalyst arrives from our previously published works. Experiments were performed in a lab-scale packed bed reactor unit that operates between 250-550 °C and 1-40 bar; for a wide range of operating conditions covering space time, ammonia and hydrogen feed partial pressures. A power law rate expression considering the effect of the proximity to the equilibrium was selected to model the reaction rate, based on the wide-spread applicability of it to the AMD process. A 1-D model was developed based on the following assumptions: (1) steady-state operation, (2) plug flow along the axial direction of the reactor with no radial variation (3) isothermal operation and (4) negligible pressure drops along the axial direction of the reactor. The experimental data was used to find the kinetic parameters: a (ammonia reaction order), b (hydrogen reaction order), k_0 (pre-exponential factor) and Ea (activation energy). A parametric estimation was performed using *fminsearch* function of Matlab®, which finds the minimum of unconstrained multivariable function using derivative-free method, setting the sum of the squared residuals between the experimental data and the calculated data as the objective function. Ammonia conversion has been used as the fitting variable. Various Matlab® subroutines were used to solve the system of differential equations: *ode45, ode23 and ode113;* with the latter offering the most promising performance. The 95% confidence intervals for all the estimated parameters were found using the *nlparci* function in Matlab®. Finally, a 3-D gPROMS model was used to further explore the effect of reactor geometry, operating conditions and catalyst geometry in the reaction performance; by using the fitted kinetic parameters.

Results and discussions

A summary of fitted parameters for both the catalysts is presented below:

Catalyst	Parameter	Value

Ru-K/CaO	k_0 (mol g^{-1} s^{-1})	(1.74 ± 0.0004) X 10^8
	Ea (kJ mol^{-1})	166.37 ± 0.0079
	a	0.47 ± 3.13 X 10^{-5}
	b	-1.42 ± 3.32 X 10^{-5}
	r^2	0.9679
	σ^2	0.096
$Co\text{-}Ba/CeO_2$	k_0 (mol g^{-1} s^{-1})	(8.55 ± 0.803) X 10^6
	Ea (kJ mol^{-1})	153.64 ± 2.72
	a	0.364 ± 0.017
	b	-1.029 ± 0.02
	r^2	0.9835
	σ^2	0.070

The 95% confidence intervals for the parameters were found to be at least three orders of magnitude lower than the parameters values, indicating high confidence in the values found. The resulting rate equation is similar to the Temkin-Pyzhev model that is widely used in the associated literature; with reaction rate being a function of positive ammonia concentration order and negative hydrogen concentration order respectively. The simulated and measured ammonia conversions from packed bed reactors for both Ru- and Co-based catalysts are in excellent agreement with each other, highlighting the validity of the model. It is predicted that the noble-metal catalyst reaches equilibrium conversion around 425 °C; which is at least 50 °C lower than the novel, cheaper Co-alternative. We further implement these parameters in the 3-D gPROMS model and find the optimal operating conditions for a catalytic packed bed reactor for both the catalysts.

Conclusions

The production of high purity H_2 from ammonia decomposition in packed bed reactor, using Ru-based and Co-based catalysts was studied theoretically and experimentally. A reliable kinetic model was implemented and validated with experimental data. The excellent agreement between modeling and experiments for both the catalysts corroborates the goodness of the developed kinetic and flow model in this work. An adequate range of operating conditions was selected to obtain high performance for both of the catalysts. Relevant information about the effect of the reactor and pellet geometries was gather for a potential scale up of the reactor.

Acknowledgements

The authors gratefully acknowledge the financial support provided by Saudi Aramco, and the resources and facilities provided by the King Abdullah University of Science and Technology.

639460 Optimizing Absorption to Improve Haber-Bosch Synthesis

Abstract:

Ammonia absorber columns offer an alternative separation unit to replace condensation in the Haber-Bosch synthesis loop. Metal halide salts can selectively separate ammonia from the reactor outlet gas mixture and incorporate it into their crystal lattice with remarkably high thermodynamic capacity. While the salts' working capacity can be limited and unstable when they are in their pure form, the capacity can be stabilized and increased by including a porous silica support.

Here, we discuss optimal conditions for uptake and release of ammonia. The production capacity (ammonia processed per unit absorbent and per unit production time) depends on processing parameters - including uptake and regeneration temperatures, ammonia partial pressures, release time, and sweep gas flow rate. Moreover, attention should be paid to balancing the release time with the full cycle time and bed size to ensure that the uptake breakthrough time makes efficient use of the bed. These parameters are mutually interdependent, so their optimization can be nontrivial, but rewarding.

To address this challenge and to allow a large set of experiments, an automated dynamic absorption system was designed and fabricated. This system was used to rapidly screen uptake and release conditions, to assess material stability, and to identify optimal cycling conditions. The large amount of experimental data obtained from this automated system, and the associated mathematical models validated by this data, were found to be an essential tool to optimize the design of the absorber column. In turn, this task is essential to assess improvement in the techno-economic prospects for this new separation approach.

We also share some of our findings regarding the scaleup. These indicate potential processing challenges that need attention in design (de-risking) to ensure optimal operation in bench-scale and production-scale units.

640180 Nuclear Hydrogen Refinery (*Pink Hydrogen to Ammonia)*

Abstract:

Nuclear Hydrogen Refinery (*Pink Hydrogen to Ammonia)*

By Richard B. Strait

Nuclear/Hydrogen Refinery: Nuclear Power plants are large water boilers producing both steam and power from the heat of the nuclear reaction. Hydrogen can be generated from electrifying water. Hydrogen on a grand scale is needed for us to move away from steam methane reforming to generate hydrogen. Wind and solar are too small to replace the SMRs we currently use but nuclear might. Nuclear power plant collocated with an industrial complex making finished products containing "green" hydrogen on an industrial scale would be a step out technology.

Fresh water supply is limited in many regions of the world and "green" hydrogen splits fresh water to produce both hydrogen and oxygen using electrolysis. Fresh water in our design comes from sea water as does most of the cooling requirement, specifically a sea water cooling tower.

The 1000 MW nuclear power plant provides the power for all applications in the generation of "green" hydrogen and those of the ammonia plant and all necessary offsites and utilities.

The ammonia plant produces 2000 MTPD of ammonia using a synthesis loop that takes it "green" hydrogen from the nuclear power plant and nitrogen from its air separation plant.

The offsites and utilities are integrated for a coastal location steam, water, "green" hydrogen and power all come from the nuclear plant. Plus port facilities for sea water intake and sea water cooling along with shipping of final products by sea.

The community and region also benefit from the supply of fresh water, power and "green" hydrogen. Many possibilities arise when excess water, power and "green" hydrogen are available.

R. B. Strait RBS1@Rice.edu

Adjunct Professor Rice University

November 30, 2021

639519 Green and Blue Hydrogen Installation Considerations

Abstract:

Hydrogen has become a much-discussed potential solution for energy suppliers and users to meet net zero carbon. A significant discussion to this point has focused on the intricacies of producing "Green" Hydrogen via electrolyzer technology to split water for electric power peak shaving and multisite, distributed applications. For higher capacity, centralized applications, "Blue" Hydrogen can be produced from traditional natural gas reforming with the addition of CO2 capture to existing facilities or inclusion in new unit design. These new and traditional process and feedstock combinations are intended to meet the objective of reducing and ultimately eliminating carbon related emissions associated with energy production AND consumption.

A next step in the evolution of Hydrogen as a fuel is bridging from technology demonstration to commercial application. To gain a better understanding of the bridging requirements, this paper will extend the Hydrogen as fuel conversation to consider the Balance of Plant or other systems needed to implement Green or Blue Hydrogen pathways. The specific areas to be covered will include feedstock characterization and handling, energy input, product use and byproduct disposition. Actual recent Green & Blue Hydrogen technology implementation and integration experience will be included.

639331 Carbon Management By Deployment of Large-Scale Otec Plants in the Gulf of Mexico

Abstract:

The world's oceans are the largest collectors and storage of solar energy. Global energy. Environment and economic settings have drastically changed: uncertain energy costs; carbon-management policy; increased energy demands; realization of impact of CO2 emissions and reorganization of Energy-Water-Nexus. This is specifically true for the Gulf of Mexico, which has been one of the favorable sites for Ocean Thermal Energy Conversion (OTEC) plants. The Gulf of Mexico is a major producer of oil and natural gas and it can be major producer of green power when oil sources decline. At current production rate of about 1.6 million barrels per day, the oil reserve will start depleting by 2040. OTEC can be the next major source of energy from the Gulf of Mexico. OTEC is 24/7 renewable energy utilizing temperature natural temperature gradients of the ocean between surface warm seawater and deep-ocean cold seawater. This energy and environment study focuses on the following key elements of large-scale deployment of OTEC plants in the Gulf of Mexico: a) production of ammonia as hydrogen carrier; b) freshwater production for the water-stressed Caribbean Islands and the Gulf States; c) conversion of captured carbon dioxide from the Gulf States to specialty chemicals; and d) recovery and conversion of carbon dioxide from seawater. The critical aspect of large-scale OTEC plants in the Gulf of Mexico is to extract the energy from the surface seawater and return the deep-ocean cold seawater at an environmentally acceptable depth. The extraction of thermal energy from the surface water and upwelling of deep-ocean cold water would reduce systematic rise of the surface water temperature of the Gulf of Mexico. This is a visionary and ambitious goal. However, in the recent years, surface temperature of the Gulf of Mexico has increased by 1-2°C, which has believed to cause severe storms with unpredictable intensity. Large-scale OTEC plants would have a positive impact on the moderation of the Gulf of Mexico seawater temperature, beside direct impact on the abatement of green-house gas emissions. Based on the previous DOE studies, an analysis was performed for production of 3,000 tonnes per day of ammonia along with desalination of seawater for assessment of techno-economic merits of large-scale OTEC plants in the Gulf of Mexico. In a similar study, techno-economic analysis was performed of utilization of captured and transported carbon dioxide from the Gulf States for conversion to methanol, dimethyl ether (DME) and alkyl carbonates. The analysis revealed that conversion of out-of-service oil platforms into OTEC platforms would significantly reduce the total-installed-cost (TIC) by taking the credit of dismantling and disposal of oil platforms. The Technology Readiness Level (TRL) of the OTEC technology is at such a stage that first of the large-scale OTEC plant can be deployed in foreseeable future with low risks.

639377 Quantifying and Comparing the Cumulative Greenhouse Gas Emissions and Financial Viability of Heavy-Duty Transportation Pathways for the Northeastern, United States

Abstract:

This study analyzes several transportation sector decarbonization pathways for the Northeastern, United States. This study calculates the financial viability in terms of net present values achieved and the cumulative greenhouse gas emissions associated with each pathway. Several commercialized Class 8 heavy-duty trucks pathways are analyzed and operate on either a standard petroleum diesel, biomass-based diesel, or a battery-electric pathway and utilize an investment timing methodology. An internalized social cost of carbon is included within the low-carbon pathways in the form of a financial benefit. Additionally, all pathways include a comprehensive sensitivity analysis. This study finds that for the baseline transportation scenarios the biomass-based diesel to battery-electric pathway is responsible for the highest NPV at $1,160,978 and is followed by the biomass-based diesel pathway at $1,155,651. In terms of greenhouse gas emissions, the biomass-based diesel to battery-electric pathway yields the lowest emissions at 5,676 megagram (Mg) CO_2e and is followed by the biomass-based diesel pathway at 7,508 Mg CO_2e.

639422 Pyrolysis and Etherification Process for Advanced Diesel Blendstocks

Abstract:

The integration of fast pyrolysis, mild hydrogenation, and etherification has been evaluated to produce a renewable diesel blendstock from woody biomass. Predictive models indicate that many ethers, which might be produced from components in pyrolysis oils, would have good blendstock characteristics, such as enhanced cetane and reduced sooting, at the expense of an energy density that is lower than petroleum diesel.

A techno-economic analysis (TEA) and a life cycle analysis (LCA) have been conducted for the integrated process. A simulation was created using Aspen Plus®, and data for fast pyrolysis at 500 °C was used for the base case. Thermodynamic properties of hydrogenation and etherification model compounds were estimated using Density Functional Theory (DFT). The results from the process simulation were used to estimate the capital and operating costs for a plant that processes 2,000 dry metric tons per day of forest residues. The well-to-wheels LCA was developed using Argonne's GREET 2019 software to assess the environmental sustainability of producing the renewable diesel fuel blendstock.

The TEA determined that diesel blendstock products with a lower heating value of 24 MJ/kg might be produced from forest residues at a minimum selling price of $2.95/ Diesel Gallon Equivalent (DGE). The requirement to remove a lower amount of oxygen from the pyrolysis oil, compared to complete hydrodeoxygenation, means that reforming of non-condensable gases from the pyrolysis reactor provides sufficient hydrogen for the mild hydrotreating reactor. Temperature effects on the pyrolysis oil composition and yields predict that the MFSP might be reduced to $2.61 per DGE at lower pyrolysis temperatures. The reduction in MFSP is due to the enriched composition of phenols and furans in pyrolysis oils produced at lower temperatures. The well-to-wheels LCA analysis indicates that the renewable diesel blendstock would have a 93% reduction in fossil fuel use and a 93% reduction in greenhouse gas emissions relative to petroleum diesel.

639427 Green Hydrogen and Renewable Natural Gas from Waste in a Circular Economy: *Producing Low-Cost Hydrogen and Rng with a Negative CO_2 Footprint from Minimally Prepared Waste.*

Abstract:

In this presentation the author describes a new approach to making green hydrogen today at less than half the cost of electrolysis and renewable natural gas at low cost, all with a low or negative CO_2 footprint. This is done by converting minimally prepared MSW, RDF, biomass, ICI, or C&D waste to a clean syngas tailored to end use, then using proven industrial processes to produce renewable hydrogen or RNG from the syngas. The negative carbon footprint results from avoiding the anaerobic decomposition of waste in landfills.

To make hydrogen, waste is first converted to syngas using OMNI's GPRS^{TM}. The syngas is then converted to hydrogen and CO_2 using a sour gas shift reactor designed and supplied by KP Engineering. Each unit will produce approximately 5000 tonnes/year of hydrogen.

The conference presentation will also describe a hybrid approach, that of combining the OMNI unit with electrolysis. In this scheme, oxygen from electrolysis is used directly in the OMNI unit, and waste heat from the OMNI unit increases the efficiency of the electrolyser. The hybrid system combines the best of both alternatives for a flexible, low-cost solution that can provide baseload hydrogen from the GPRS^{TM} and peaking operation from the electrolyser, with significantly better economics than the electrolyser alone.

RNG is made by hydrogenation of the gas using mature industrial processes and/or in conjunction with electrolysis, using the oxygen for the gasifier and the hydrogen to adjust the H_2:CO ratio.

The process guarantee is backed by an insurance wrap and supported by data from over a decade of demonstration at a commercial scale. Delivery is de-risked by delivery in 200 tpd shop assembled modular units and choices made to ensure high availability.

639576 Cleaner Energy: Fossil Fuels Finds Forward Thinking

Abstract:

Despite some amazing progress of the emerging energy tech that has surfaced in the last number of years, the fact remains that our dependence on finished petroleum products for manufactured-goods and energy, is with us for the foreseeable future. (*McKinsey and Company, 2021)*

While similar comments can be found in other expert articles supporting the intensive, industrial complexes surrounding the US Gulf Coast, it doesn't mean that oil refineries and chemical plants can continue to operate as they've done in the past.

In order to remain competitive, US Gulf Coast processors will need to adopt new technologies and methods, in their pursuit of both profit margins and favorable public opinion. To remain relevant, and be seen as part of the clean energy future, these processors need to demonstrate a mindset for employing new technologies that reduce the energy these plants consume and the emissions they create.

One of the leading culprits for energy consumption in these Coastal plants, are the thousands of *'heat exchangers'* required by these processors for the refining of the raw materials. The key challenge with the various forms of heat exchange equipment such as: boilers; chillers; evaporators; condensers and heaters—is simply, that they become dirty or fouled.

The annual loss attributable to heat exchanger fouling in the US and UK together is $16.5 billion. Heat exchanger fouling has a direct impact on plant profitability and is one of the costliest problems facing the refining and chemical industries today.

Heat exchanger fouling can cause a wide range of problems to occur if action is not taken to periodically remove deposits or pre-emptively, prevent deposit-formation in the first place. For the purposes of this paper, we focus our presentation on two of the more notable problems:

1. Fouled heat exchangers require far more energy to operate, Vs., when they are clean,
2. CO_2 emissions increase exponentially in fouled heat exchangers if left unresolved.

In most industrial situations, the inefficiency of heat transfer resulting from fouling has a direct link to excess fuel consumption in the process. In addition to the appropriate selection of operating conditions and exchanger geometry, there are numerous chemical and mechanical methods to mitigate fouling and to remove deposits from the heat transfer surfaces.

However, all methods to reduce fouling require some understanding of the mechanisms of the deposition process, the structure of the deposits and the factors which govern adhesion to the heat transfer surfaces. Even with optimal design and operation, cleaning of the heat exchanger surfaces eventually becomes necessary.

Having to clean is by no means a new phenomenon. In fact, this dirty equipment continues to foster a multi-billion dollar collective of industrial-services companies, dedicated to this annual expenditure by US Gulf Coast refiners & processors.

What if there was a technology that significantly reduced heat exchanger fouling and in some cases eliminated the cleaning cycle for years?

This paper will reveal the results from a study conducted by one of the major, US Gulf Coast refiners as they adopted a specific fouling mitigation technology in Europe, to reduce energy consumption within one of their processing units.

Empirical data shared from the 16 month trial, demonstrates that specific goals of overall fouling reduction and increased *'run-length'*, were achieved for this particular refiner. However, it was the energy recovery portion and reduction in process emissions, that topped the operational benefits. The annual energy recovery was documented at US $1.4million for this one processing unit.

Don't mistake this paper for just another technology vendor hawking their wares. We chose to draw your attention to heat exchanger fouling, because it's a big beast. Are there other waste culprits? Sure. But consider this: One of, if not the largest, petrochemical refiners in the Gulf Coast has processing facilities in both Beaumont and Baytown (TX). Collectively, they will clean over 2500 heat exchangers annually, due to fouling. How much energy could be recovered, if they didn't have to clean?

Achieving cleaner energy is indeed possible. Exploring this future through cleaner processing equipment, via new technologies, is just one of the ways Gulf Coast processors can ensure themselves of being part of the sustainable, profitable and forward thinking industries in the region.

AUTHOR INDEX